Carlos Henrique Alencar Almeida
Cicero Souto

Thermoelectricity direct generation

Carlos Henrique Alencar Almeida
Cicero Souto

Thermoelectricity direct generation

An automated experimental analysis

ScienciaScripts

Imprint

Any brand names and product names mentioned in this book are subject to trademark, brand or patent protection and are trademarks or registered trademarks of their respective holders. The use of brand names, product names, common names, trade names, product descriptions etc. even without a particular marking in this work is in no way to be construed to mean that such names may be regarded as unrestricted in respect of trademark and brand protection legislation and could thus be used by anyone.

Cover image: www.ingimage.com

This book is a translation from the original published under ISBN 978-620-2-17539-5.

Publisher:
Sciencia Scripts
is a trademark of
Dodo Books Indian Ocean Ltd. and OmniScriptum S.R.L publishing group

120 High Road, East Finchley, London, N2 9ED, United Kingdom
Str. Armeneasca 28/1, office 1, Chisinau MD-2012, Republic of Moldova, Europe
Printed at: see last page
ISBN: 978-620-8-08977-1

Table of contents:

To my wife Jéssica for her understanding and
patience in waiting, listening and advising...
To my brothers and my father for believing in my
potential...
To my mother (*in memorlum*) for her inspiration...
To my son Nicolas to inspire him,

Dedication.

ACKNOWLEDGMENTS

The author would like to thank the Federal University of Paraíba's Laboratory of Active Systems and Structures (LaSEA), the National Council for Science and Technological Development (CNPq) for their financial support and the Federal Institute of Paraíba for the logistics required for the conciliation.

To those who helped carry out these studies.

To my coworkers for their facilitation, changing schedules, changing subjects, applying activities when necessary.

To Inakan, Kléber, Edleuson, Diego and Regina for their kind accommodation in times of need.

To Pierre, Robério and the Nóbrega family for their company, logistics and meaningful friendship.

To José, Bruno, Celso, André, Adriano, Alexsandro and other research colleagues for sharing their knowledge.

To the advisor for making this possible, for closely following every step of the evolution of this work and meeting the needs that arose.

Finally, and most importantly, to God for enabling me to overcome difficulties, tiredness and distance, for presenting me with the best conditions and opportunities and for the wisdom to make the right decisions and find the answers I needed.

SUMMARY

CHARACTERIZATION OF A THERMOELECTRIC CELL FOR ELECTRICITY GENERATION

This work characterizes a thermoelectric device as a generator of electrical energy. For this purpose, a thermoelectric cell was subjected to different temperature profiles on its faces. Heat was applied to the device through a feedback control structure made up of cascaded Peltier cells, temperature sensors and current conditioning circuits. The system was run and monitored via a data acquisition and control interface connected to a computer and managed by *software* dedicated to the application. With these experiments, the electrical voltage response of the thermoelectric device was observed in relation to the temperatures applied to its faces. A resistive load was also inserted to analyze the behavior of the electrical power supplied by the device. Among the results obtained, the variation in the Seebeck coefficient when the average working temperature is varied stands out. The power supplied by the thermoelectric device configured as a generator reaches 95 mW when subjected to a temperature difference (AT) of 40 °C. The power and current curves are shown in relation to the voltage generated by the device.

Keywords:thermoelectricity,   energy generation electricity, energy harvesting, alternative energy.

Chapter 1

1 INTRODUCTION

The growing demand for electricity in recent years, especially with regard to the use of fossil fuels as raw materials to produce it, has led to rising global temperatures, natural disasters and high levels of toxicity. CO2 concentration levels already exceed 6 billion tons (BROWN, 2003; MELBOURNE, 2010), responsible for the greenhouse effect that contributes to the increase in global temperature.

In 1998, the United Nations Framework Convention on Climate Change drew up the Kyoto Protocol (C&T, 1998) which, among other efficiency issues, deals with reducing the production of toxic gases released into the atmosphere. In this document, Item "2.1.(a).(i)" deals with increasing the energy efficiency of relevant sectors of the national economy. Faced with this reality, countries have been working to reduce the production of CO2 released into the atmosphere. Among the strategies that the various sectors have been using to adapt to better conditions, the energy generation sector has sought to pay attention to forms of electricity generation that are more sustainable, based on classic concepts (McDONOUGH, 2002).

Among the various forms of sustainable generation, priority has been given to techniques that use renewable resources such as the sun, wind or other natural phenomena. In Australia, for example, several research and development institutes have launched the Beyond Zero Emission plan (BZE, 2013), which renews concepts and policies for the generation, distribution and consumption of electricity. Based on this new concept, the BZE launched the Repowering Port Augusta campaign, which aims to replace two coal-fired power stations, Playford B (240MW) and Northern (520MW), with renewable energy stations. Among the proposals in this program are thermoelectric plants that use the sun as a heat source.

This concentrating solar power generation technology is already being applied in various plants, such as the PS10 and PS20 towers in Seville, Spain, manufactured by Abengoa Solar, for example (GOOD, 1996). Some developed countries are already well advanced in this context, making extensive use of energy from photovoltaic plants or other renewable sources.

One of these countries is Germany, which today has around 25% of its population living in regions with 100% of its population living in these regions. powered by renewable energy, which was considered almost a revolution. The United States is home to the largest photovoltaic power plants today, including the Agua Caliente Solar Project in Arizona, USA. These systems for generating electricity from the sun now have a generation capacity and efficiency such that they can compete with classic generation sources in terms of sustainability standards (SILVA, 2012).

The generation of electrical energy from heat accounts for almost all of the world's electrical matrix (EPE, 2012). Among the most common forms, heat is used to produce movement, which is the main phenomenon 0. However, this is not the only way to generate energy through heat. As an alternative to the classic forms, the principle of thermoelectricity (ATTIVISSIMO, 2014; ENGELKE, 2010; BOBEAN, 2012; BOBEAN, 2013; DISSALVO, 1999) is a source to be explored, which can increase the possibilities of energy production, as well as on different value scales.

The use of thermoelectricity to generate electricity, combined with the forms of heat collection and storage currently in use, suggests an excellent source of research for a new form of electricity generation, which could be even more sustainable due to the possible reduction in CO_2 production, but with reservations regarding production capacity.

Based on this research, this paper presents the characterization of a thermoelectric cell (Peltier module) as an electricity generator. This study was made possible through experimental analysis of the system in various situations, using an analysis structure capable of relating the power generated to the behavior of the temperature on its faces.

For this characterization, the cell was subjected to temperature variations on both sides. The potential difference at the device's terminals was captured in real time and compared to the temperatures to which the device was subjected, as well as the difference between them. An analysis was also made of its behaviour as an electricity generator, adding a resistive load to the system.

1.1 OBJECTIVE

General objective:

The aim of this work is to characterize a thermoelectric cell as a generator of electrical energy in response to controlled temperature variations on its faces, with a view to low-power electronic applications.

Specific objectives:

- Determine the static and dynamic behavior of the thermoelectric cell's electricity production;

- Study the capacity to generate electrical voltage;

- Determine the cell's internal impedance;

- Study the capacity to produce electrical power.

Chapter 2

2 LITERATURE REVIEW

The use of thermoelectricity associated with other effects can be exemplified by PAUL (2014), where the author associates thermoelectricity with the photovoltaic effect, making an analysis based on a mathematical model of the structure, considering the sun's radiance in the locations analyzed. The aim of the work was to improve the efficiency of electricity generation, concluding that performance was better in areas with higher irradiance or closer to the Equator.

ENESCU (2014) reviews the performance parameters of a thermoelectric device as a thermal machine, determining its figure of merit, cooling capacity and coefficient of performance. It also analyzes the incidence of correlated effects, such as the impact of the Tomson effect. It also traces electrical characteristics, such as the impedance or electrical resistance of the device.

To better understand the Peltier effect in thermoelectric devices, HAIDAR (2008) suggests a cascade association of two modules, where the system can be analyzed separately using the equivalent electrical circuit. In this condition, the device in the first stage reaches a temperature difference as expected due to its characteristics, but it passes a reduced temperature to the device in the second stage on its hottest side, so that the cold side can reach even lower temperatures. The model can be understood as shown in Fig. 1.

FIG. 1. CASCADE ASSOCIATION OF PELTIER CELLS (ADAPTED FROM HAIDAR 2008).

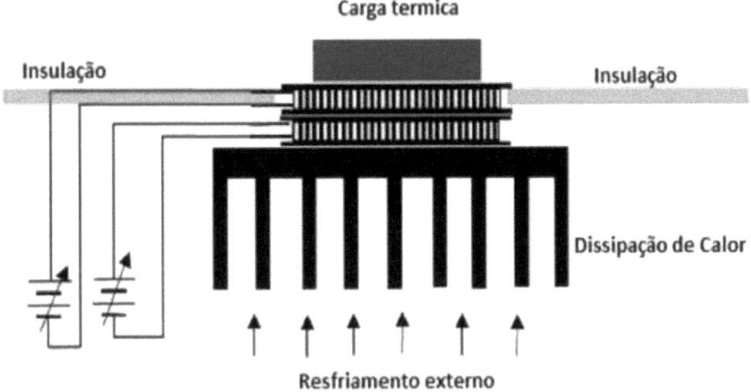

In the research by SOUTO (2014), the phenomenon of thermoelectricity is applied under a cascade association of Peltier modules to control the temperature on the surface of an SMA/PZT heterostructure, making it possible to vary the temperature from 4°C to 96°C.

KYLAN (2010) analyzes the Seebeck effect in various metal alloys, some of which have excess electrons (N-type) and others excess gaps (P-type), subjecting them to various temperature and pressure conditions, finding their physical limits and Seebeck coefficients. A prototype cell was then made with an alternating combination of P-type and N-type materials under a structure that favored heat flow. This prototype was then coupled to the exhaust of a combustion-powered electric generator, taking advantage of the heat produced in the generation of rotary motion to generate thermoelectric energy in order to increase efficiency.

The possibility of using thermal energy to generate electricity through thermoelectricity was analyzed by FERNANDES (2012), where the author used Peltier cells coupled around a chimney, receiving the heat dissipated on one side and keeping the other side at room temperature. His proposed structure used 80 Peltier cells measuring 1 square inch, capable of producing a power output of 208 kWh. It is worth noting that in the case analyzed, the financial return on the proposed investment was approximately 143 years, based on the values of the resources used at the time of development.

A thermoelectric device (TG 100) exposed to temperature differences was the case analyzed by BOBEAN (2013). The author designed a structure consisting of an electrical resistor and a ventilated heat sink, where a point-by-point analysis was carried out using voltmeters, relating the temperatures presented on the hot and cold sides, the time elapsed and the electromotive force at the terminals of the thermoelectric device, reaching a voltage of approximately 4.5V at a maximum temperature of 116°C, with a difference between the temperatures of 30°C.

The scientific contribution of SCHAEVITZ (2001) was the use of joining different materials under a car exhaust. The author used a polycrystalline silicon-germanium structure doped with phosphorus to obtain the P-type semiconductor and another doped with boron to obtain the N-type. His device was able to work stably at temperatures of up to 500°C with voltages of up to 7 volts and an efficiency of 0.02 %.

Some patents propose ways of using the temperature of other processes to produce energy, such as HANSON's patent (1975) which proposes a thermoelectric device made up of P-type and N-type semiconductor junctions positioned in the exhaust system of a motor vehicle.

KUMMER's patent (1969) proposes a structure for subjecting a metal junction to differences in pressure and temperature, causing a flow of ions to appear under an electrode.

OJEDA (2015) used the PSO (particle swarm optimization) algorithm to obtain parameters for thermoelectric modules. The parameters obtained were the thermal capacitance, thermal conductance and temperature of each part of a structure made up of two thermoelectric devices thermally coupled by a layer of aluminum, as well as fins and fans for dissipating heat to the environment. In comparison with real conditions, the mathematical model obtained by this article obtained a similar behavior in the graph of temperature as a function of time.

GYÖRKE (2015) uses a photovoltaic solar panel associated with a thermoelectric device to power wireless sensors. The behavior of the device throughout the day was analyzed, taking into account changes in brightness. With a difference of only 8 °C and a radiance of 10 W/m^2 , it was possible to achieve powers of over 900 uW/cm^3 (microwatts per cubic centimeter).

AJIWIGUNA (2015) determines the figure of merit of a thermoelectric device by relating thermal resistance, electrical resistance and Seebeck coefficient, obtained through a measuring system according to Harman's method. The results show that the Seebeck coefficient of the device increases as the average temperature rises. It also concludes that the highest temperature experienced causes the lowest thermal resistance.

VERAS (2015) also determines the figure of merit of a thermoelectric device using the Harman method. To obtain these results, a test platform was used, capable of applying a specific thermal cycle. A degradation in performance was observed over 548 thermal cycles, where before the cycles the figure of merit was 89.4×10^{-3} and after the cycles it was 72.6×10^{-3} .

ALMEIDA (2015) developed a thermal test structure capable of printing different temperature behaviors against the faces of a thermoelectric device in order to determine its performance as a voltage source in relation to the temperature difference. The results were stored in real time and the relationship between the temperature difference and the voltage produced was determined according to the Seebeck equation.

After this review, it was decided to focus this work on characterizing the thermoelectric device as an electricity generator, relating the input energy source to the amount of electrical power delivered to a given load. The behavior of the device was analyzed constantly, providing instantaneous information throughout the experiments, thus presenting the real behavior of the device through variations in the temperature difference.

Chapter 3

3 THEORETICAL BACKGROUND

3.1 PELTIER EFFECT

Thermoelectricity is divided into three phenomena: Peltier, Seebeck and Thompson. Discovered by Frenchman Jean Charles Athanase Peltier in 1834, the Peltier phenomenon suggests the presence of a temperature gradient at the junctions between two different conductors in an electrical circuit, when traversed by an electric current. Considering a junction point between two different conductors, one type-p and the other type-n, when a continuous electric current is applied to their terminals, the flow of electrons will produce a concentration of thermal energy at one end followed by a release of energy at the other. As a result, both elements will conduct heat from the junction point, making it colder than the other end (Fig. 2A) (KASAP, 2001; LINEYKIN, 2007; DISALVO, 1999).

FIG. 2 A-THE PELTIER EFFECT; B-THE SEEBECK EFFECT (ADAPTED FROM DISSALVO, 1999).

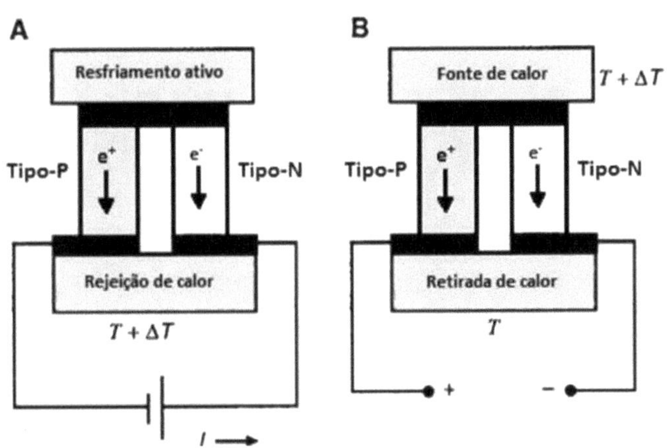

3.2 SEEBECK EFFECT

In 1821, the Estonian Thomas Johann Seebeck presented the first definitions of the principle of thermoelectricity, relating a present temperature gradient in a bimetallic structure to a magnetic field present in its vicinity, which was noticed by some change in the polarity indicated by a compass where, by relating it to the research of his contemporary Oerested, he deduced that an electric current was flowing in the conductors of his structure (TEJEDOR, 2006).

FIG. 3 SEEBECK APPARATUS (FRANCE, 1920)

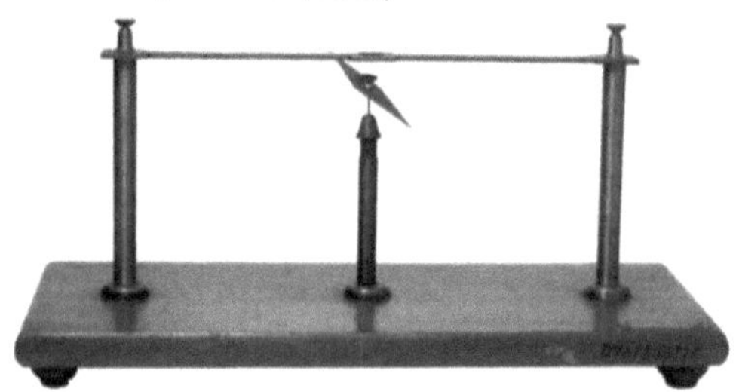

Photo: Paulo Noronha Filho (2010)

Still considering the structure described in Fig. 2B, according to the explanation of the phenomenon, when the junction is heated, heat will flow to the other end, causing a potential difference to appear at the terminals of the colder end. This is due to the tendency of electrons to move away from the hotter side towards the colder side. Therefore, as the materials are different, there will be a difference in the amount of electron displacement, presenting itself as a potential difference (SOUZA, 2013; ENGELKE, 2010; BOBEAN 2012; BOBEAN, 2013).

Each material behaves differently to these phenomena. Therefore, the quantification of the phenomenon is called the Seebeck Coefficient, in which the relationship between the quantities involved can be represented by equation (1).

$$\alpha = \frac{U}{T1 - T2}$$

(1)

Where a is the Seebeck Coefficient, U is the voltage at the device terminals, T1 is the Temperature at the junction and T2 - Temperature at the end

(FERNANDES, 2012).

Among the various materials, metal alloys and semiconductors have the best coefficients when associated with platinum (BOBEAN, 2013).

According to research carried out in the mid-twentieth century, the semiconductors silicon and germanium, when doped, can present a Seebeck effect with similar behavior and a voltage response corresponding to the doping (P-type with positive voltage and N-type with negative voltage). Among these, Silicon shows a better response to the effect when it is at temperatures between 400 and 500K (~120 to 220°C), while Germanium shows its best response at a temperature around 300K (26°C), with a lower voltage response than Silicon (GEBALLE, 1954; GEBALLE 1955).

The combination of doped semiconductors is very favorable for thermoelectricity, as suggested by Thompson's research in the mid-19th century, but in small dimensions (TEJEDOR, 2006). To obtain a considerable effect, an association of small P-N junctions is proposed, as described in Fig. 2 and Fig. 4, thus forming two thermal surfaces, as suggested in INOUE's patent (1998).

FIG. 4 PELTIER MODULE, ACCORDING TO PATENT US 5841064 A OF 1998

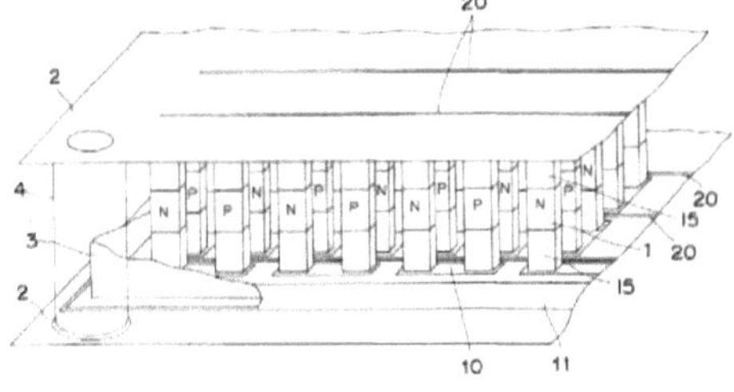

Commercially, a module is used, made up of multiple silicon P-N junctions that form two surfaces as shown in Fig. 5. For better thermal conduction, the commercial module also has a ceramic layer on each side. However, the application of the Seebeck effect to metallic alloys has not been ruled out, as their behavior is fairly linear and they can withstand higher temperatures than semiconductors.

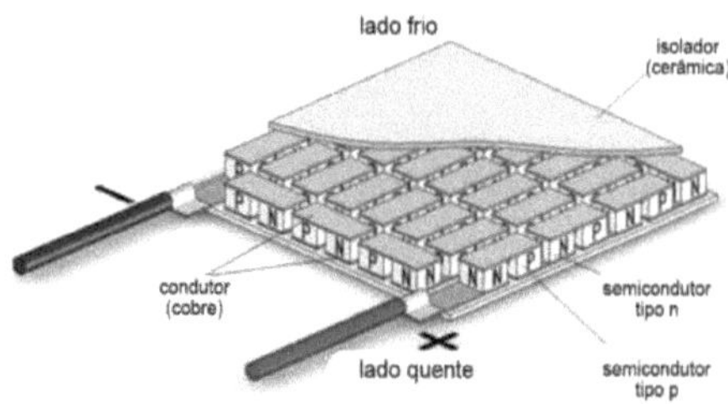

FIG. 5 COMMERCIAL PELTIER CELL

In this case, the combination of different metal alloys is widely used as temperature sensors (thermocouples), requiring electronic conditioning for a better reading of the voltage generated by the temperature difference.

Chapter 4

4 MATERIALS AND METHODS

In order to achieve the objectives, a structure was designed which is capable of controlling the temperature of both sides of the device under analysis by manipulating the voltage values in its control circuits. The structure has sensors for system feedback and real-time data analysis via computer.

Here is a detailed description of each item.

4.1 THE MODULE

Due to its greater availability on the market, the thermoelectric module (Peltier module) made up of multiple silicon-based P-N junctions was chosen for this research, Fig. 6. Based on the data provided by the manufacturer, the results obtained in this work and the results of the research by VERAS (2015), who used the same module as this research, the following characteristics can be highlighted:

- Manufacturer: Hebei I.T.
- Surface: Ceramic (Al2O3)
- Composition: BiSn
- Number of PN junctions: 127
- Dimensions: 40 mm x 40 mm x 4 mm
- Maximum operating temperature: 138 °C
- Maximum current: 5.3 A
- Maximum voltage: 16.2 V

- Ohmic resistance: $2.4 \sim 2.75\,\Omega$
- Seebeck coefficient: $23 \sim 36$ mV/°C
- Electrical conductance: $0.368\,\Omega/mm$
- Figure of merit: $72.6 \times 10^{-3} \sim 89.4 \times 10^{-3}$

- Thermal conductivity: 2 K.mm/mW

FIG. 6 PELTIER MODULE

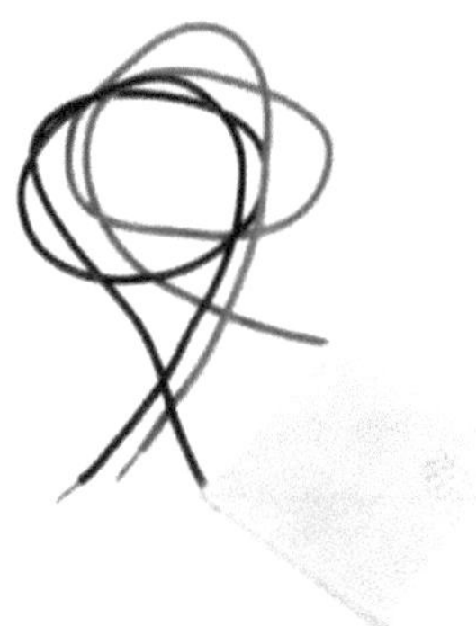

This module's power paths are marked with red and black, because in the application for which it is intended, the direction of the current in its terminals determines which side will be heated and which will be cooled, and there is reciprocity when this direction is reversed.

4.2   ACQUISITION SYSTEM

To monitor, analyze and control the variables in question, some physical equipment was used to acquire data and some software to control, analyze and process the information.

Hardware:

- The DAQ NI-USB-6008 was used for data acquisition with analog inputs of ±10 Volts with a resolution of 12 bits at a rate of 10,000 samples per second;

- The NI9213 thermocouple input module used two inputs configured for fast reading. This module has a cold junction compensation system and automatic zeroing to compensate for offset errors. It reads in 24 bits and has a sensitivity of up to 0.02 °C at an update rate of 75 samples per second;

- Two ports of the NI9263 output module with 16-bit resolution and ±10 Volt signal and 100,000 samples per second update (Fig. 7);

Software:

- LabVIEW™ is a graphical environment for developing systems with complete and specific interaction for each of the devices used;

- Matlab® for data processing and graph plotting.

FIG. 7 (LEFT) NI USB-6008 USED FOR VOLTAGE READING; NI9213 FOR TEMPERATURE READING; (RIGHT) NI9263 FOR VOLTAGE OUTPUTS. (COURTESY OF NATIONAL INSTRUMENTS ®).

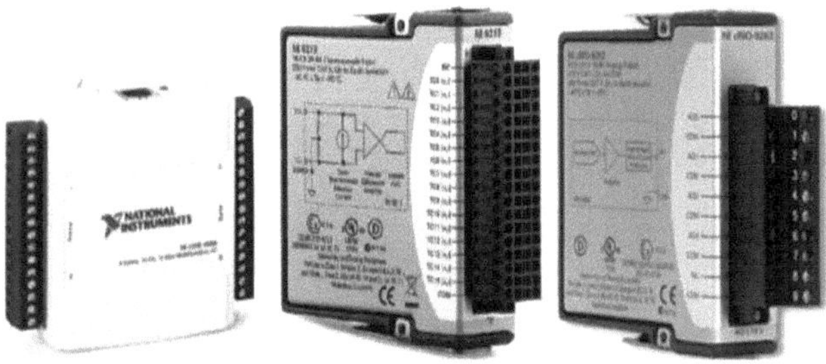

In this *software*, an interface has been designed in which the user can change the temperature values, the analysis time, the number of repeat cycles, the maximum and minimum temperature values, the temperature difference between hot and cold or even the voltage values on the analog outputs, as well as monitoring the temperature and voltage values generated in real time. These values are then delivered to an algorithm which controls the process and all the information is stored in a database.

4.3 CONTROL STRATEGY

Proportional-Integral-Derivative control was used to achieve the desired temperature values on the faces of the device under analysis. The variables of this control were:

- reference signal (*setpoint*): desired temperature;

- signal transduction: current control circuit;

- output signal: temperature;

- feedback: thermocouples.

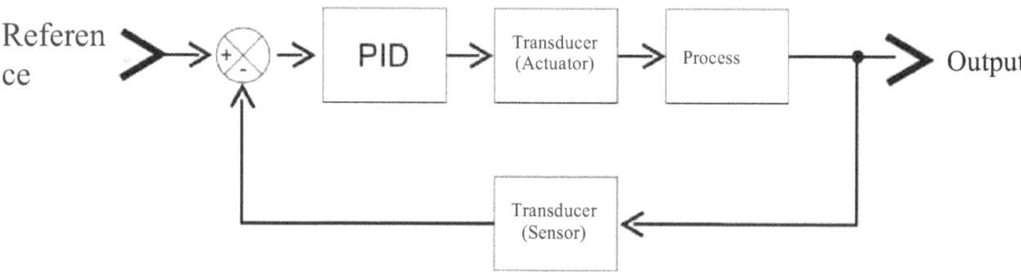

The block diagram in Fig. 8 shows the relationship between the different quantities involved in this system, where the desired temperature and the control setting are information obtained from the computer and converted into an electrical signal by the DAQ. The temperature of the system changes in proportion to the current current applied to the heating/cooling devices. The thermocouples, in turn, translate the temperature information into a voltage signal, returning to the system the information needed for feedback to the controller.

The system receives the desired temperature information from the user and goes through the virtual control block, then through signal transduction until it reaches the output temperature variation. This value, however, is exposed to variations in the ambient temperature. To ensure that the reference temperature is reached, it is necessary for a reading of the output temperature to be compared to the desired value, thus characterizing a closed-loop system, similar to BOLTON's model (1995).

For temperature control, the PID controller has a better response when low values are assigned to the derivative constant (DORF, 2001; NISE, 2007; OGATA, 1970), due to the system's long response time. In a case of temperature control by pulse width adjustment (ARRUDA, 2014), a very low value was assigned to the derivative constant, whereas in this work, the value assigned to this constant was zero, and the controller can only be considered PI (proportional-integral).

In this system, the control response is close to instantaneous, since the speed of reading and writing updates to the DAQ's analog ports, as well as the processing of information, is much greater than the response time of the temperature at its output.

The same diagram is applied separately to each of the temperature control units.

4.4 REFERENCE SIGNAL

The "REFERENCE" variable of the system described above in Fig. 8 can be fed into the system with the maximum and minimum temperature values chosen by the user. This value in turn is delivered to the system with a behavior that varies over time. For the purposes of this work, the behavior corresponding to the following classic analysis waveforms was adopted:

- Step;
- Ramp;
- Sine.

These waveforms can be assigned to each of the control systems separately, but for this experiment, the following combinations were adopted:
- Set the temperature on one side close to the ambient temperature while oscillating the temperature on the other side according to the waveform;
- Oscillate both temperatures with the same waveform, maintaining a constant temperature difference between them.

4.5 TEMPERATURE CONTROL

The temperature control, as shown in the block diagram above, consists of a computer-controlled feedback thermoelectric structure.

FEEDING

To monitor the temperature, a thermocouple was used on each of the faces, positioned in the middle of a thin sheet of copper between the faces of the module under analysis and the heating/cooling module (Fig. 9).

FIG. 9. THERMOCOUPLE HOUSING.

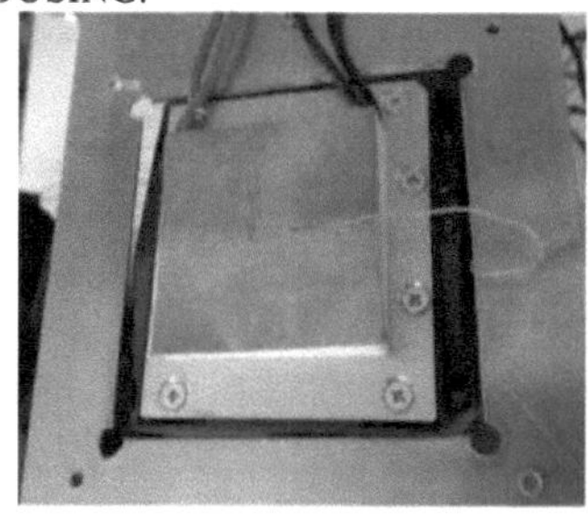

TEMPERATURE ADJUSTMENT

To adjust the temperature, two structures similar to the one described by HAIDAR (2008) were built. In each of the structures, an active heat sink with exhaust and thermal fluid circulation is used, as well as two electronically controlled Peltier modules associated in cascade, covered in Styrofoam. In each of the structures it is possible to reach temperatures below 0°C or above 100°C (Fig. 10).

FIG. 10. ANALYSIS STRUCTURE

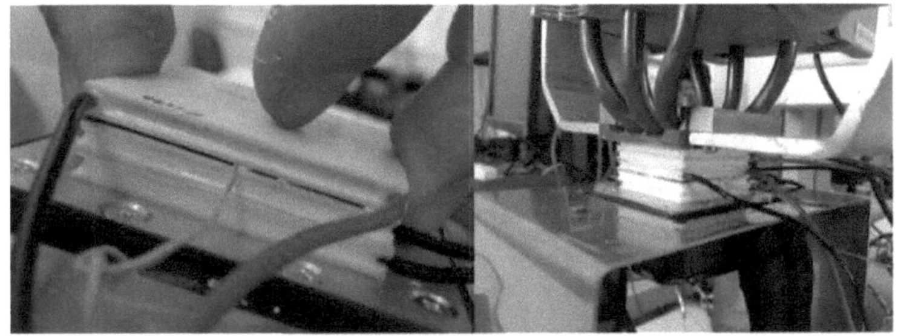

To control the temperature in the devices mentioned, the system designed manipulates the analog outputs of the DAQ, delivering the necessary signal proportionally to their respective current control circuits.

4.6 ASSEMBLING THE STRUCTURE

The diagram in Fig. 11 describes the assembly of the analysis bench where, for each face of the module, there is a thermocouple positioned in the center of a thin, heat-uniforming copper foil. After the foil, there is a cooling/heating structure made up of two cascaded Peltier modules and an active heat sink (as described above). The system's Peltier module assembly is protected by a layer of polyethylene (Styrofoam) to reduce the thermal effects of the environment. A thin layer of thermal paste is applied between each layer of the assembly. Two current control circuits are connected to the Peltier pairs for individual cooling/heating control.

FIG. 11. BENCH ASSEMBLY DIAGRAM. A-COOLING SYSTEM; B-MODULE UNDER ANALYSIS; C-POLYIMIDE TAPE; D-TEMPERATURE SENSORS.

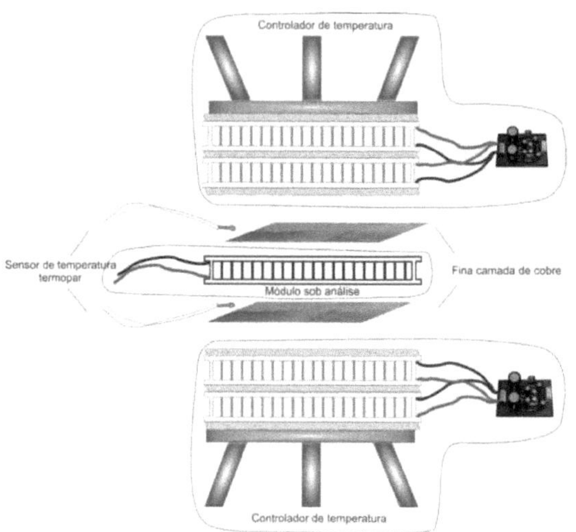

The thermocouple terminals are connected to the NI 9213 Temperature Acquisition Module, the input ports of the two current control circuits are connected to the NI 9263 Analog Output Module and the terminals of the module under analysis are connected to the NI 6008 acquisition device. The three interfaces are connected to the computer via USB 2.0 and their information managed by the LabVIEW ® *software* application.

4.7   THERMOCOUPLE CALIBRATION

To calibrate the thermocouples connected to the NI 9213, an Agilent U1242® multimeter was used in the temperature meter position with a resolution of 0.1 °C. The two thermocouples to be calibrated and the reference thermocouple are positioned on the face of the thermal source, joined and fixed with a specific adhesive tape for applications of varying temperatures.

FIG. 12. CALIBRATION PROCEDURE.

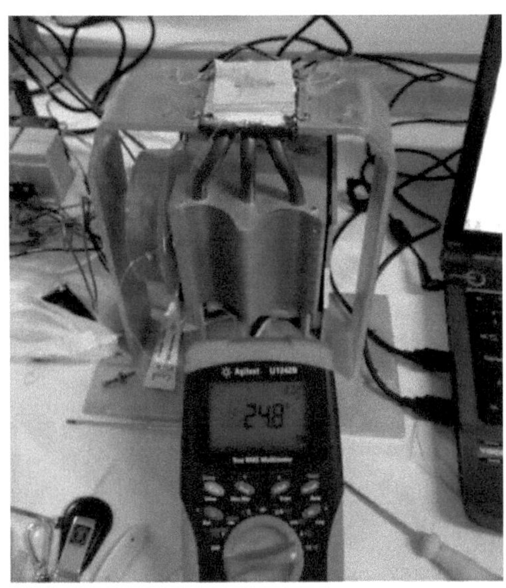

The procedure involves applying an electric current to the thermoelectric devices (Peltier module), changing the temperature on the analysis surface. After the time required for the temperature to stabilize, the reference in the calibration system is filled in with the value shown on the multimeter display. This is followed by confirmation of the entry of that point in the calibration table, saving the instantaneous information of the thermocouple under calibration.

The calibration values of the two thermocouples used in the experiment can be seen in Table 1.

TABLE 1 CALIBRATION POINTS

Date	29/04/2015				
Reference -20,0	T1 (°C) Measurement -20,00	Error (R - M) 0	Reference -20,0	T2 (°C) Measurement -20,00	Error (R - M) 0
-8,2	-8,04	-0,16	-14,5	-9,09	-5,41
-6,0	-5,26	-0,74	-12,0	-7,17	-4,83
-4,5	-4,11	-0,39	-9,0	-4,48	-4,52
-3,0	-2,60	-0,4	-8,0	-4,13	-3,87
-2,0	-1,71	-0,29	-6,0	-2,52	-3,48

-0,3	-0,44	0,14	-4,0	-1,02	-2,98
0,8	1,12	-0,32	-2,0	-0,20	-1,8
1,3	1,77	-0,47	0,0	2,48	-2,48
2,0	2,41	-0,41	2,0	4,91	-2,91
4,0	4,07	-0,07	4,0	5,91	-1,91
6,5	6,80	-0,3	6,0	8,41	-2,41
8,0	8,18	-0,18	8,0	9,50	-1,5
10,5	10,89	-0,39	10,0	11,04	-1,04
13,0	13,39	-0,39	13,0	13,71	-0,71
16,0	16,27	-0,27	16,0	16,06	-0,06
18,5	18,64	-0,14	19,0	18,85	0,15
21,0	20,93	0,07	23,3	23,15	0,15
23,0	22,88	0,12	25,0	24,57	0,43
25,0	25,20	-0,2	25,3	25,14	0,16
28,0	28,21	-0,21	28,0	27,53	0,47
31,0	30,72	0,28	29,8	29,31	0,49
35,0	34,49	0,51	32,0	31,29	0,71
40,0	39,44	0,56	33,0	32,38	0,62
45,0	44,58	0,42	36,0	35,22	0,78
50,0	49,13	0,87	40,0	39,37	0,63
55,0	53,91	1,09	46,0	45,18	0,82
60,0	58,74	1,26	50,0	48,90	1,1
65,0	64,13	0,87	55,0	53,80	
70,0	68,94	1,06	60,0	58,28	
75,7	74,91	0,79	66,0	64,01	
80,0	78,99	1,01	71,0	68,96	
85,0	84,10	0,9	76,9	74,93	
90,0	90,18	-0,18	81,0	79,37	
98,0	97,76	0,24	88,0	85,69	
103,0	102,53	0,47	95,0	92,18	

The curves generated by these values are shown in Fig. 13, showing good linearity in the region between 20 °C and 80 °C, where the experiments in this research were concentrated.

FIG. 13 THERMOCOUPLE CALIBRATION CURVES

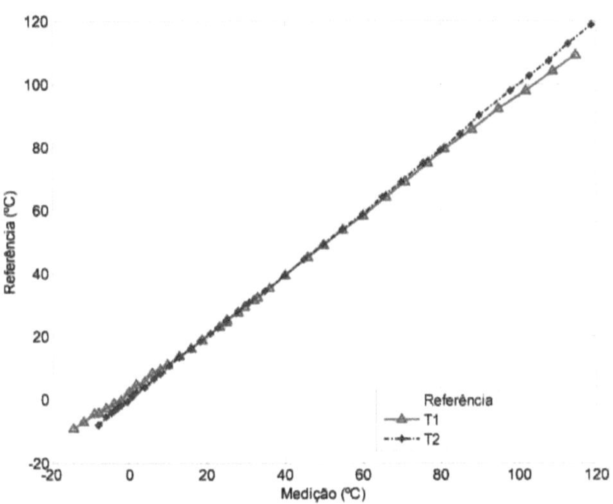

4.8 INTERFACE

An interface (Fig. 14) was developed using LabVIEW® software to manipulate and observe the system's variables in real time.

In this interface, the user can manipulate certain variables to comply with the desired measurement strategy. The variables include the maximum and minimum temperature of each face or the fixed difference between the two temperatures. You can also choose which temperature will remain fixed so that the other temperature varies according to the waveform adopted, or even choose for both to oscillate together, maintaining a temperature difference chosen by the user.

FIG. 14. ANALYSIS INTERFACE. A-VALUE OF THE TEMPERATURE OF BOTH FACES; B-MEASUREMENT OF THE STRESS GENERATED; C-DEMANDED VALUE FOR THE TEMPERATURES.

Fig. 14 shows the interface during operation, at the exact moment when the temperature equivalent to T1 in the Seebeck equation (1) is approximately 73

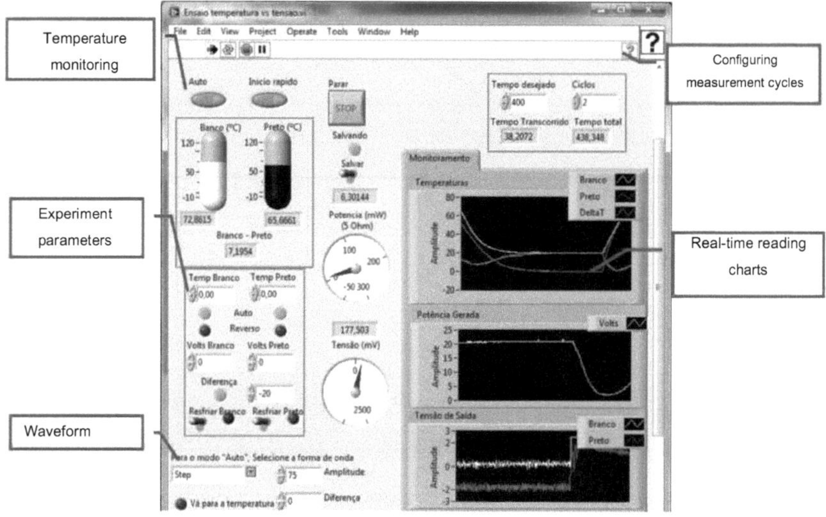

°C and the temperature T2 is approximately 66 °C ("Temperature monitoring"). At this point, the voltage at the terminals of the module under analysis is almost 177 mV. The two sides are set to automatic mode ("Experiment parameters"), but with a difference of 20 °C between them. The waveform adopted for the experiment was Step ("Waveform >>step"), with an amplitude of 75 °C. In "Configuration of measurement cycles", 2 (two) cycles of 400 s were requested. The moment of the screenshot was at 38 s of the 2nd cycle (Total time 438 s).

The graphs shown in the "Monitoring" tab are, in the first window from top to bottom ("Temperatures"), the temperatures on the faces of the module under analysis and the difference between them, in the middle window ("Power generated"), the graph of the power after a sudden reduction in the difference between the temperatures, in the other window ("Output voltage") are the voltages at the analog output of the NI 9263 module, limited to ±2.5 V. Note that the moment in the screenshot points to the moment when the voltages at the output of the interface rise in response to a sudden rise in temperature.

The block diagram for programming this interface can be seen in Appendix I.

Chapter 5

5 RESULTS AND DISCUSSIONS

5 **RESULTS AND DISCUSSIONS**

To analyze the electrical energy generated in the cell under analysis, the system was subjected to different situations, applying temperature behaviors corresponding to the waveforms described in chapter 4.4. The behavior of the voltage at the module terminals was analyzed a *priori, with* no load, analyzing the thermoelectric effect. A resistive load was then added to determine the power behavior in relation to the temperature behavior.

5.1 THE CHOICE OF CARGO

Based on the Maximum Power Transfer theorem (BOYLESTAD, 2004), we chose a resistance value capable of halving the value of the open voltage (high impedance). The value of the resistor capable of this was approximately 5fi, so an ohmic resistor with this value was adopted for the power-related experiments. The graph (b) in Fig. 15 shows that the voltage doubles in value when the load is removed (5fi) and returns to its initial value when the load is reinserted.

FIG. 15 TEMPERATURE (A) AND VOLTAGE (B) BEHAVIOR DURING INSERTION/REMOVAL OF A 5 OHM LOAD.

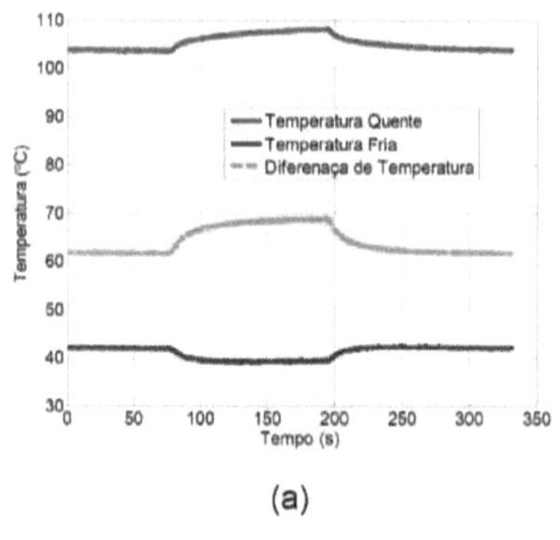

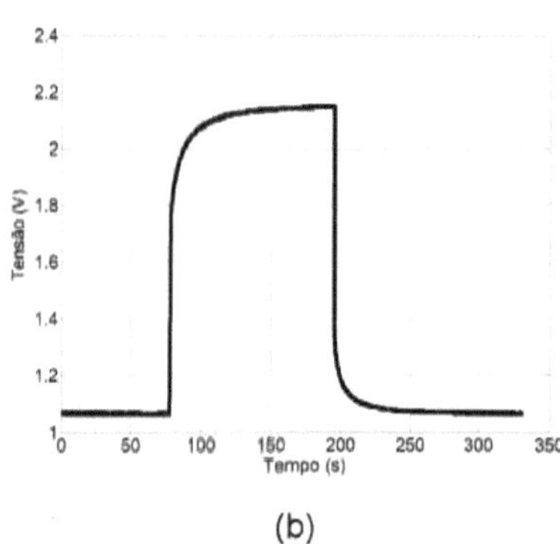

(a)

(b)

Another important fact to note in Fig. 15 is the behavior of the temperatures on both sides. When the load is inserted, the temperature difference stabilizes at 60 °C. However, when the load is removed, this difference increases symmetrically to 70 °C, i.e. the hot side gets hotter and the cold side gets colder. This phenomenon is based on a possible reciprocal effect in the passage of electric current (when there is a charge), such as the Peltier effect described above. In this case, the direction of the current determines the opposite temperatures (cooling the hot side and heating the cold side).

5.2   SETTING THE TEMPERATURE ON ONE SIDE

In this experiment, a temperature of 20 °C is set on one side and maintained until the end of the experiment. An oscillating temperature is applied to the other side, according to the wave adopted, from 20 °C to a peak of 60 °C, thus obtaining a difference between the temperatures of up to 40 °C.

For each waveform, readings were taken with and without load. The algebraic difference between the temperatures was represented in a third curve on each graph.

In all three situations, the system was able to keep the cold-side temperature static until the complete cycle of the waveform adopted for the hot-side temperature.

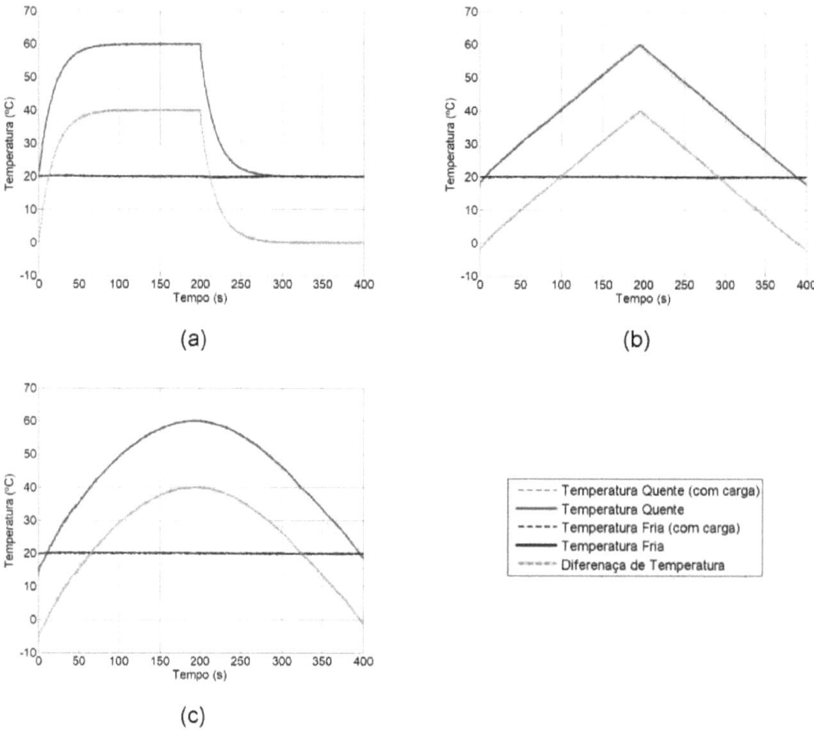

Fig. 16 STABILIZING THE TEMPERATURE ON THE COLD SIDE AT 20°C AND RAISING THE TEMPERATURE ON THE HOT SIDE TO 60°C. WAVEFORMS: STEP (A), TRIANGULAR (B) AND SINUSOIDAL (C)

Fig. 17 shows the behavior of the electrical voltage without load and the electrical voltage with load (dashed) with their values on the vertical axis on the left. On the right are the values of the line that represents the behavior of the power over the load. The horizontal axis represents the time elapsed during the experiment.

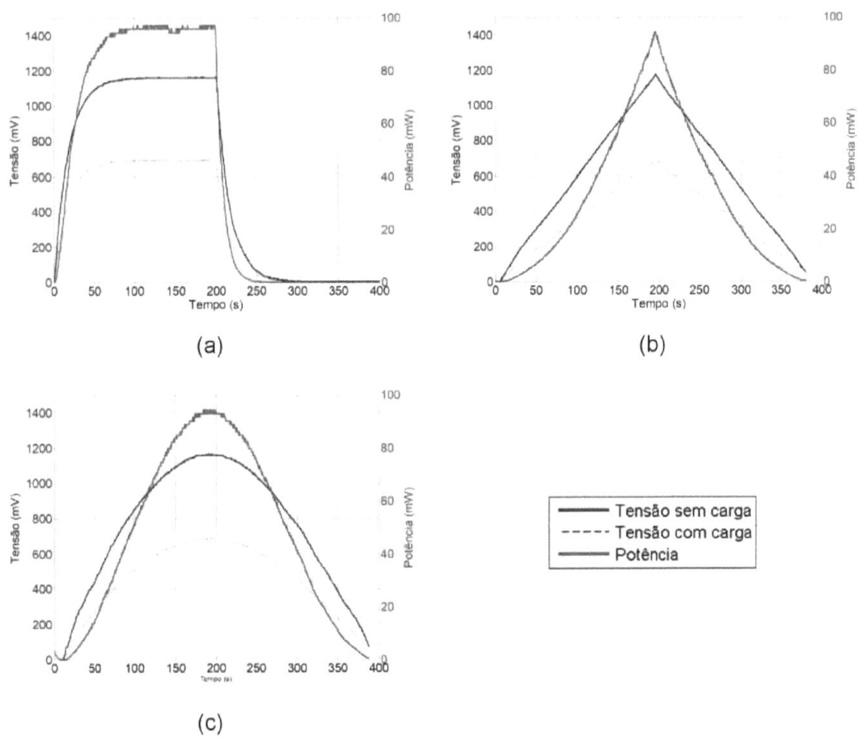

Fig. 17 VOLTAGE AND POWER RESPONSE. WAVEFORMS: STEP (A), TRIANGULAR (B) AND SINUSOIDAL (C)

It can be seen that in both cases the power value reaches a peak of approximately 95 mW in the same periods in which their temperature graphs reach differences of 40 °C. This result is consistent with the relationship described by the Seebeck equation, where the voltage value must be proportional to the temperature difference between the junctions. As an ohmic resistance is being used, the power response should also be proportional to the temperature difference. However, the behavior of the power is different from that of the voltage as the temperature difference rises.

For a better understanding of the behavior of each quantity, the graphs in Fig. 18 relate the temperature difference to the no-load voltage and the power (with load) for each waveform.

FIG. 18 VOLTAGE AND POWER AS A FUNCTION OF AT. WAVEFORMS: STEP (A),
TRIANGULAR (B) AND SINUSOIDAL (C)

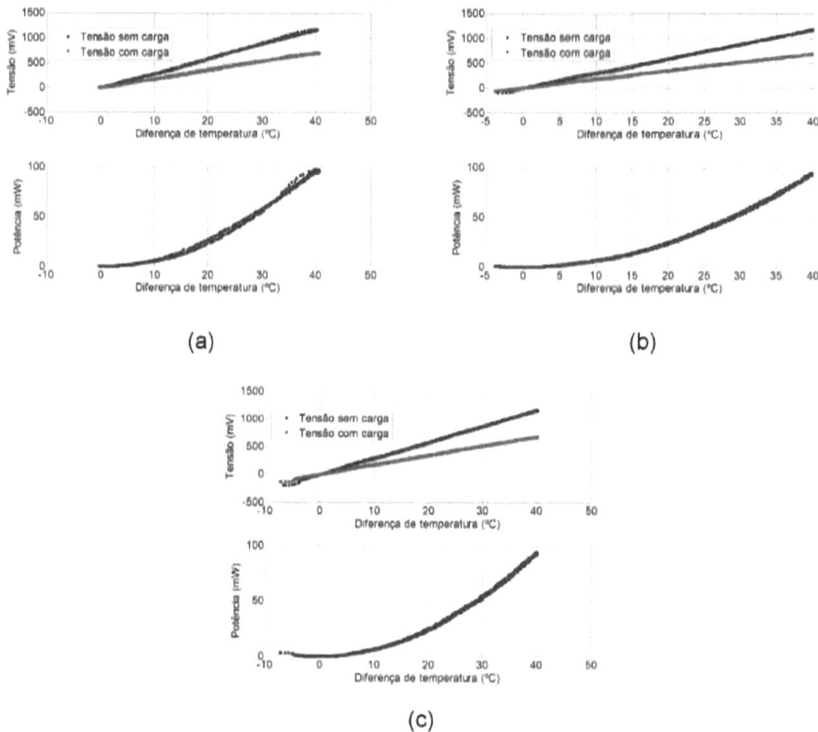

(a)

(b)

(c)

The curve of the no-load voltage in relation to the temperature difference behaves like a straight line, as predicted by the Seebeck equation. In the case of power, the curve shows non-linear behavior, but no energy storage.

5.3 FIXING A DIFFERENCE OF 20 °C BETWEEN THE TEMPERATURES OF THE FACES, WITH THE HIGHER ONE OSCILLATING UP TO 95 °C

The Seebeck equation relates the voltage at the terminals of the thermoelectric device to the temperature difference between the faces and also to the coefficient of the material. This test analyzes the possibility of this relationship varying when the TA is moved at different ranges and under different behaviors.

Trying to maintain a constant difference of 20 °C between the temperatures on the faces of the device, these temperatures were subjected to different waveforms, observing the system's response to each one.

In fig. 19, when implementing step (a), there were moments when the TA could not be maintained, but this difference returned to the desired value after a few seconds. This is due to the slow response of the temperature to a sudden change in reference.

FIG. 19 MAINTAINING A FIXED DIFFERENCE BETWEEN THE FACES OF 20 °C AND OSCILLATING THE WARMER ONE UP TO 95 °C. WAVEFORMS: STEP (A), TRIANGULAR (B) AND SINUSOIDAL (C)

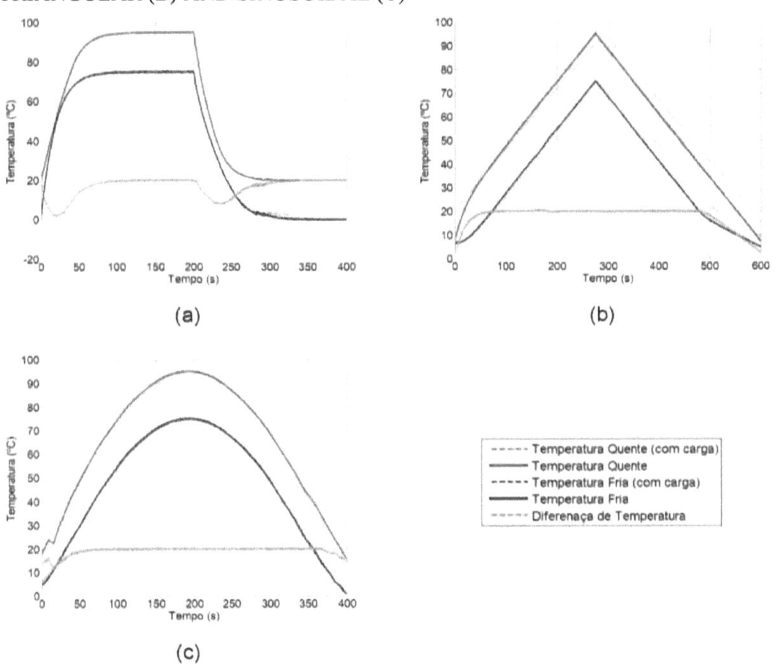

For this situation, what can be seen is that the behavior of the no-load voltage and the power was neither similar nor proportional to ΔT. In the case of step (a) this fact is more easily perceived, where in the period from 50 to 200 seconds the no-load voltage value, for example, stabilizes at around 700 mV, and after the period of instability (from 200 to 300 seconds), this value drops to around 500 mV, while the HV graph (fig. 19 (a)) returns to the same value (20 °C).

These two different voltage values when subjected to the same AT indicate that there was a change in the Seebeck coefficient of the device when subjected to different temperature regions.

FIG. 20 RESPONSE TO THE LINEAR BEHAVIOR OF AT IN DIFFERENT TEMPERATURE REGIONS.WAVEFORMS: STEP (A), TRIANGULAR (B) AND SINUSOIDAL (C)

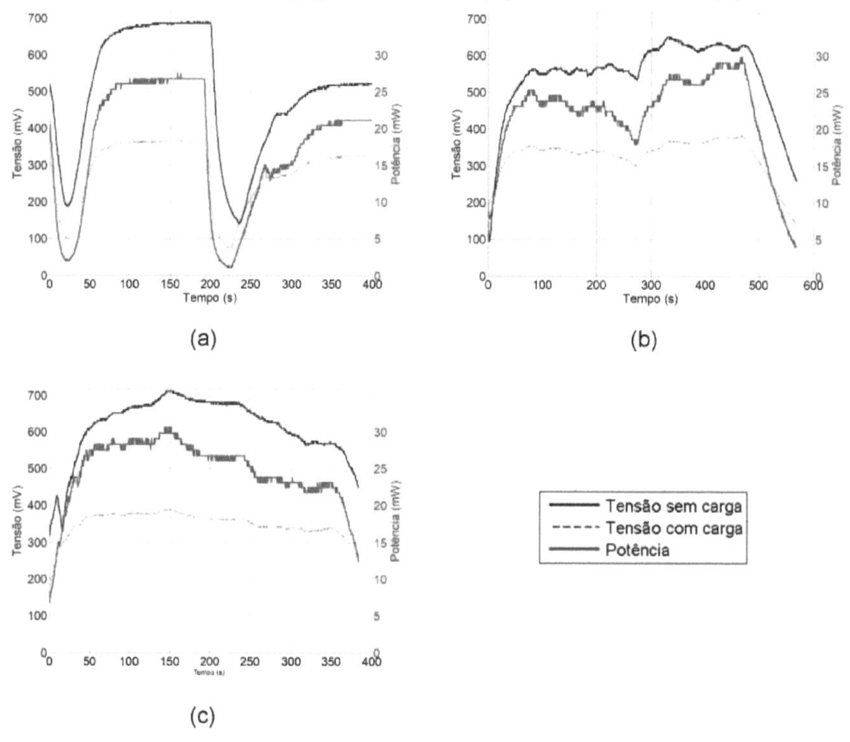

This is repeated in the case of ramp (b) where, during the sudden change of direction in the temperature curves, even without there being a change in the HV curve, there was a drop in the power and voltage curves. After this point (300 s), the voltage and power values were slightly higher (450 s) than when the temperature values were the same, but rising (250 s).

With sine (b), the experiment was more stable, even so, the voltage and power curves (Fig. 20 (b)) showed variations in values. At the beginning, the electrical voltage and power values were slightly higher than at the same temperature values, but they decreased. In other words, between 50 s and 100 s the no-load voltage was above 600 mV, while between 300 s and 350 s it was below 600 mV. During both periods, the HV value remained stable at 20 °C (fig. 19).

The change in the value of the coefficient is easily seen when you remember that the Seebeck equation relates the stress to the multiplication of the coefficient by the AT. The latter having remained stable, the change in the stress value is then attributed to the change in the Seebeck coefficient

Analyzing the Voltage vs AT relationship in the case of the step (Fig. 21 (a)), the temperature difference is maintained until half the time in a region above 70 °C and after half the time below 20 °C. The curve of this relationship shows two distinct regions that concentrate more points, (in addition to the scattering of points in other regions due to the moment of transition between one step and another). These two regions clearly point to values close to 500 mV and 700 mV, with a very large disparity in relation to other regions.

FIG. 21 VOLTAGE AND POWER AS A FUNCTION OF _. FOR CONSTANT AT. STEP (A), TRIANGULAR (B) AND SINE (C) WAVEFORMS

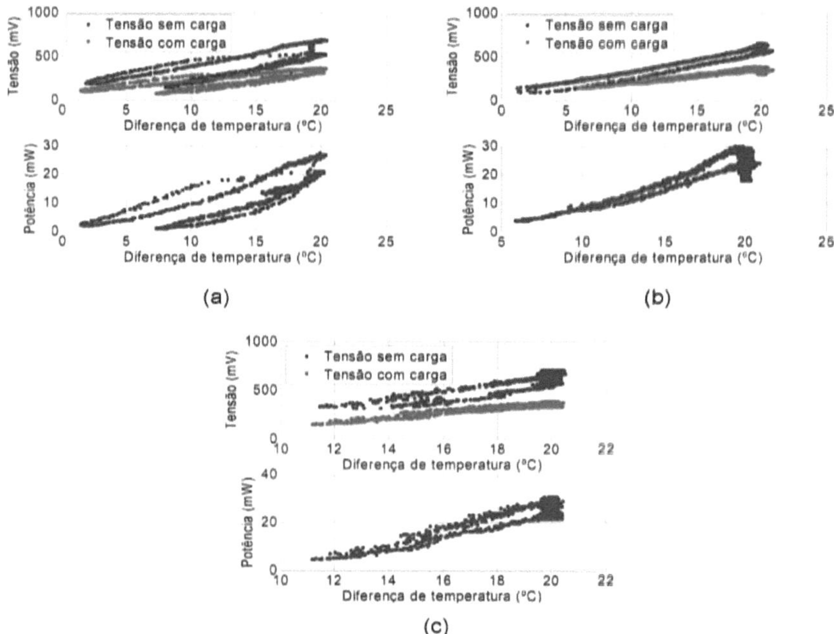

For the triangular and sinusoidal forms, the accumulation of points in the analysis is more concentrated, since the change from one temperature region to another is not as abrupt as in the step. The greatest accumulation of points, in both cases, occurs in the regions with the highest power and voltage values, thus showing a similarity between their behaviors, as seen in the respective graphs.

In the power graphs, the points are only scattered on the ΔT axis within a range of 1 °C, which can be considered as just one point, but with a scatter of almost 10 mW on the Power axis. This indicates that the power does not only depend on ΔT, but also on the temperature range in which this value is found.

5.4 SEEBECK COEFFICIENT

As described by AJIWIGUNA (2015), the Seebeck coefficient varies with the average temperature. This means, for example, that the ΔT of a temperature region with a minimum of 20 °C and a maximum of 60 °C is the same as the ΔT where the minimum is 40 °C and the maximum 80 °C, but because the average temperatures are different, the coefficient is also different.

Fig. 22 shows the behavior of the Seebeck coefficient in relation to the average temperature of this study, obtained using Equation (1).

The average temperature is determined by the sum of the minimum temperature and the difference between the maximum and minimum, as shown in Equation (2).

$$T_{media} = T_{frio} + \Delta T \tag{2}$$

In Fig. 22 (a), the situation mentioned is the one shown previously in Fig. 16 (a), where the temperature on the cold side is fixed and the temperature on the hot side oscillates. This is a case in which there is no change in the temperature region, since the temperature on the cold side remains constant. In this case, the Seebeck coefficient remains constant. The same occurs in cases (c) and (e).

Fig. 22 (b) corresponds to the situation described in Fig. 19 (a), where the temperature difference remains fixed and the temperature on both sides fluctuates. In this case, there is a change in the temperature region where ΔT is located, with the coldest region having a minimum of 0 °C and a maximum of 20 °C and the hottest region having a minimum of 75 °C and a maximum of 95 °C. The same situation occurs in (d) and (f). For these cases, the Seebeck coefficient does not remain constant, and the graph shows two prominent regions.

FIG. 22   RELATION OF THE SEEBECK COEFFICIENT ΛT WITH THE AVERAGE TEMPERATURE. (A) STEP WITH _ OSCILLATING; (B) STEP WITH ΛT FIXED, (C) TRIANGULAR WITH AT OSCILLATING; (D) TRIANGULAR WITH ΔT FIXED; (E) SINUSOIDAL WITH ΔT OSCILLATING; (F) SINUSOIDAL WITH ΔT FIXED.

In some situations where ΔT is kept fixed by changing the temperature region, there were times when the value of the temperature difference approached zero due to the limitation of the system, bringing the seebeck coefficient to infinity. These

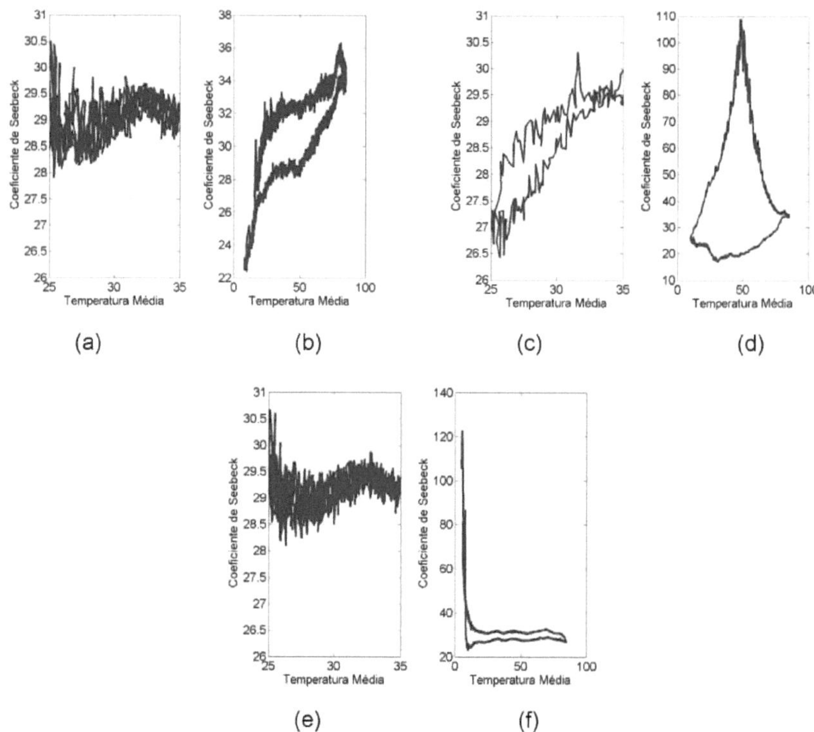

periods should be disregarded from the analysis.

With the exception of transients, the values obtained for the seebeck coefficient (α) ranged from a minimum of 23 to a maximum of 36 (mV/°C). This result points to greater electrical power in higher temperature regions.

5.5 THERMOELECTRIC GENERATOR POWER CURVE

When analyzing this device as an electricity generator, it is necessary to relate its ability to produce voltage to its ability to supply current, thus determining its behavior as such.

Maintaining a constant temperature difference and controlled in a closed loop, an adjustable resistive load was applied to the terminals of the thermoelectric device. The voltage and current values were then read off for different resistance values, thus obtaining the graph in Fig. 23. This procedure was carried out for ΔT of 25 °C and 30 °C. The load resistance was varied from 1 to 30 Ω.

The table with the measurement values ranges from a Short Circuit to a resistance of $30\ \Omega$, at the rate of $1\ \Omega$.

The highest power point is around 150 mW for a ΔT of 40 °C. Compared to photovoltaic systems, this is a reasonable value, since a photovoltaic panel with an area 1000 times larger produces around 250 W in an ideal situation.

FIG. 23 POWER CURVE (A), MEASURED POINTS (B).

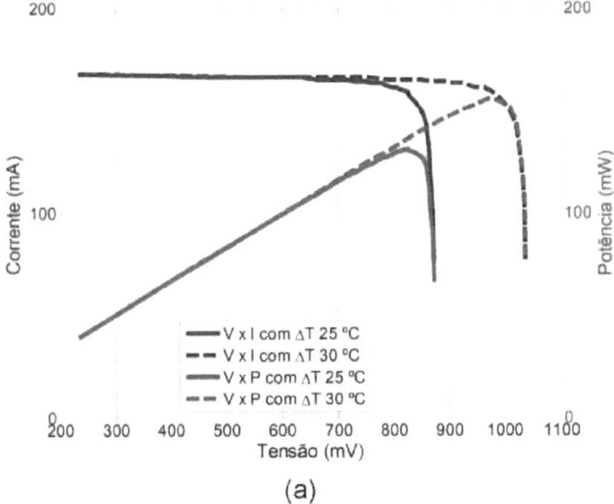

(a)

ΔT = 30 °C		
R (Ω)	I (mA)	V (mV)
1	168	267
2	167	429

	I (mA)	V (mV)
3	166	550
4	166	636
5	166	705
6	166	752
7	165	791
8	165	821
9	165	847
10	164	870
11	164	891
12	163	909
13	163	924
14	162	937
15	161	947
16	161	962
17	160	971
18	159	977
19	158	982
20	156	990
21	154	997
22	152	1002
23	150	1009
24	147	1013
25	143	1019
26	138	1022
27	132	1025
28	121	1030
29	105	1034
30	76	1036

$\Delta T = 26\ ^{\circ}C$

R (Ω)	I (mA)	V (mV)
1	168	234
2	167	375
3	166	469
4	166	539
5	166	589
6	166	633
7	165	664
8	165	689
9	165	712
10	164	733
11	164	749
12	163	760
13	163	777
14	162	788
15	161	797
16	161	805
17	160	815
18	159	825
19	158	826
20	156	832
21	154	839
22	152	844
23	150	849
24	147	854
25	143	859
26	138	860
27	132	863
28	121	865
29	105	870
30	76	872

(b)

5.6 FINAL CONSIDERATIONS

In this work, the behavior of the voltage and electrical power of a thermoelectric device was analyzed when subjected to temperature variations.

These analyses were carried out with the controlled application of different combinations of temperature waveforms on the faces of the device, capturing their response by a real-time data acquisition and control system. For the power analysis cases, a resistive load of 5 Ω was used, chosen on the basis of the maximum power transfer rule.

The response of the phenomenon, specifically the generation of voltage and power in the thermoelectric module studied, was evaluated up to the temperature difference that the thermoelectric device was capable of producing, not exposing its thermoelectric limitations.

Among the conditions to which the device under analysis was exposed, the greatest temperature difference reached on its faces was approximately 70 °C without load and almost 60 °C with load.The potential difference at the device terminals was 2.2 V and 1.1 V, respectively.In the experiments with the application of heat curves, the temperature difference that the system could control was 40 °C, which was the limit used to analyze the behavior. For this value of ΔT, the voltage obtained was 1.2 V.

With regard to the power delivered to the resistive load of 5 Ω inserted in the system, an exponential behavior can be seen in the graph Power vs. ΔT. As the temperature variations
the temperature difference reached 40 °C with a power delivered to the load of 95 mW.
In the most extreme case, with a ΔT of 60 °C, the power was approximately
242 mW.

It was observed that when a load is added to the thermoelectric generator, the value of ΔT. Esta decreases.
In other words, the hotter side gets colder and the colder side gets hotter. This reinforces the fact that the difference in temperature is the driving force behind the thermoelectric generator.

It was observed that there is a variation in the Seebeck coefficient when the average temperature is shifted to different regions, i.e. from 0 - 20 °C the coefficient is different from that of 10 - 30 °C, for example, both with $\Delta T = 20$ °C.

Finally, given the voltage and power values provided by the thermoelectric module analyzed, it can be said that they can be used to power low-power devices, especially wireless sensor systems.

CONCLUSION

This work was based on the proposal to characterize a thermoelectric device as an electricity generator, relating the temperature on its faces to the voltage and power produced.

Based on the results obtained, it is considered possible to use a thermoelectric device to generate electricity on different scales, using any temperature gradient, as well as a combination of units.

The electrical power found in the experiment has a relevant value compared to other alternative systems for generating electricity, such as a photovoltaic system, for example.

It is interesting to note that due to the simplicity of the thermoelectricity phenomenon, it can be used to power low-consumption embedded systems or even to generate cogeneration in large systems. A 1600 mm device^2 is capable of producing approximately 100 mW with a temperature difference of 40°C , which can be even higher
as the temperature difference increases. Compared to other forms of generation that use heat (such as solar towers), when analyzing the generation capacity per square meter, in proportional terms, thermoelectricity could reach even higher values.

With this information comes the demand for investment in more research in this area, with a view to a new, possibly more sustainable form of electricity generation.

FUTURE WORK

Based on the conclusions of this study, future work is proposed:

- The choice of other thermoelectric devices to compare them with the one studied in this work;

- Study the combination of several thermoelectric modules to produce more electrical power;

- Make a comparison of its generation capacity with other forms of electricity generation that have the concept of harvesting electricity;

- Build a prototype for harvesting electricity from heat sources, such as the sun, using thermoelectric cells;

- Test the use of electrical energy produced by thermoelectricity in commercial low-power systems.

REFERENCES

AJIWIGUNA, T. A., et al. "Measurement System for Thermoelectric Module" In: Instrumentation and Measurement Technology Conference (I2MTC) Proceedings, 2015 IEEE International. IEEE, 2015. p. 1889-1893.

ALMEIDA, C. H. A. et al. "Characterization of thermoelectric cell for electric power generation". In: Instrumentation and Measurement Technology Conference (I2MTC) Proceedings, 2015 IEEE International. IEEE, 2015. p. 1358-1362.

ARRUDA, Bruno Willian de Souza; LIMA, Robson Pacífico Guimarães; SOUZA, CleonilsonProtásio De. An artificial immune system-based anomaly detection method applied on a temperature control system. International Journal of Industrial Electronics and Drives, v. 1, n. 3, p. 145-153, 2014.

ATTIVISSIMO, F. et al. Photovoltaic-thermoelectric modules: a feasibility study. In: Instrumentation and Measurement Technology Conference (I2MTC) Proceedings, 2014 IEEE International. IEEE, 2014. p. 659-664.

BOBEAN, Crina; PAVEL, Valentina. The study and modeling of a thermoelectric generator module. In: Advanced Topics in Electrical Engineering (ATEE), 2013 8th International Symposium on. IEEE, 2013. p. 1-4.

BOBEAN, Eng Crina; PAVEL, Valentina.DIDACTICAL STAND FOR THE STUDY OF THERMOELECTRIC GENERATORS, 2012

BOLTON, William. Control engineering. Makron Books, 1995.

BOYLESTAD, Robert L. Introduction to circuit analysis. Pearson Education do Brasil, 2004.

BROWN, Lester R. "Eco-economics: building an economy for the earth. "Salvador:

Uma (2003)

C&T Brasil; Kyoto Protocol, Translated by the Ministry of Science and Technology, 1998;

CHEN, Yu-Chieh et al. Temperature sensing and controlling biological experiments by using one thermoelectric module. In: Instrumentation and Measurement Technology Conference (I2MTC) Proceedings, 2014 IEEE International. IEEE, 2014. p. 1314-1317.

Rio de Janeiro, Empresa de Pesquisa Energetica. Statistical Yearbook of Electricity 2013. 2012.

DISALVO, Francis J. Thermoelectric cooling and power generation. Science, v. 285, n. 5428, p. 703-706, 1999.

DORF, Richard C.; BISHOP, Robert H. Modern control systems. Technical and Scientific Books, 2001.

ENESCU, Diana; VIRJOGHE, Elena Otilia. A review on thermoelectric cooling parameters and performance. **Renewable and Sustainable Energy Reviews**, v. 38, p. 903-916, 2014.

ENGELKE, Kylan Wynn et al. Novel thermoelectric generator for stationary power waste heat recovery. 2010. Doctoral dissertation. Montana State University-Bozeman, College of Engineering.

FERNANDES, Alberto Emanuel Simões dos Santos. Energy conversion with Peltier cells. 2012. Available at: http://hdl.handle.net/10362/8084

GEBALLE, T. H.; HULL, G. W. Seebeck effect in germanium. Physical Review, v. 94, n. 5, p. 1134, 1954.

GEBALLE, T. H.; HULL, G. W. Seebeck effect in silicon. Physical Review, v. 98, n.

4, p. 940, 1955.

GOOD, Brian S.; CHUBB, Donald L.; LOWE, Roland A. Optimization study of selective emitter thermophotovoltaic systems. In: Energy Conversion Engineering Conference, 1996. IECEC 96, Proceedings of the 31st Intersociety. IEEE, 1996. p. 1007-1012.

GYÖRKE, P. ,PATAKI, B. "Energy harvesting wireless sensors for smart home Applications In: Instrumentation and Measurement Technology Conference (I2MTC) Proceedings, 2015 IEEE International. IEEE, 2015. p. 1757-1762.

HAIDAR, S.; ISAAC, I.; SINGLETON, T. Thermoelectric Cooling Using Peltier Cells in Cascade. 2008.

HANSON, Charles M. Thermoelectric voltage generator. U.S. Patent n. 4,095,998, June 20, 1978.

HERRMANN, Ulf; KELLY, Bruce; PRICE, Henry. Two-tank molten salt storage for parabolic trough solar power plants. Energy, v. 29, n. 5, p. 883-893, 2004.

http://kryothermtec.com/portugal.html. Accessed on 30/10/2014

http://ofrioquevemdosol.blogspot.com.br/2013/05/alemanha-transicao-or-revolucao.html - Accessed on Thursday, May 2, 2013.

http://www.abengoasolar.com/web/en/nuestros_productos/plantas_sol ares/, Accessed May 2, 2013;

INOUE, Hiroyuki et al. Peltier module. U.S. Patent no. 5,841,064, Nov. 24, 1998.

KASAP,   Safa.   Thermoelectriceffectsinmetals   :
thermocouples. Canada: Department of Electrical Engineering
University of Saskatchewan, 2001.

KUMMER, Joseph T. Thermo-electric generator. U.S. Patent n. 3,458,356, July 29,

1969.

LINEYKIN, Simon; BEN-YAAKOV, Shmuel. Modeling and Analysis of Thermoelectric Modules. IEEE TRANSACTIONS ON INDUSTRY APPLICATIONS, v. 43, n. 2, p. 505, 2007.

McDONOUGH, W.; BRAUNGART, M.; Design for the Triple Top Line: New Tools for Sustainable Commerce Corporate Environmental Strategy, Vol. 9, No. 3, Elsevier Science Inc.

Melbourne Australian sustainable energy: Zero carbon Australia stationary energy plan. University of Melbourne, 2010.

National Instruments Corporation; USER GUIDE SPECIFICATION, N. I. USB 6008/6009. , 2012.

NISE, Norman S. CONTROL SYSTEMS ENGINEERING, (With CD). John Wiley & Sons, 2007.

OGATA, Katsuhiko; YANG, Yanjuan. Modern control engineering. 1970.

OJEDA, D. et al. "Parameter Identification of Thermoelectric Modules using Particle Swarm Optimization". In: Instrumentation and Measurement Technology Conference (I2MTC) Proceedings, 2015 IEEE International. IEEE, 2015. p. 812-817.

RepoweringPort Augusta - BZE,available at http://bze.org.au/, accessed in May 2015;

SCHAEVITZ, Samuel B. et al. A combustion-based MEMS thermoelectric power generator. In: The 11th International Conference on Solid-State Sensors and Actuators. 2001. p. 30-33.

SOUTO, Cicero da R. et al. Thermal cycling effect on a shape memory and piezoelectric heterostructure. In: Instrumentation and Measurement Technology

Conference (I2MTC) Proceedings, 2014 IEEE International. IEEE, 2014. p. 315-319.

SOUZA, Ricardo Farias de. CHARACTERIZATION OF THE SEEBECK EFFECT IN HETROGENIC COPPER OXIDE JOINTS, in VIII EPCC - Encontro Internacional de Produção Científica Cesumar, ISBN 978-85-8084-603-4, Maringá - Paraná - Brasil, 2013.

SILVA, M. R., "Um Modelo para Análise da Sustentabilidade de Fontes Elétricas", UFPE, Recife-PB, 2012.

TEJEDOR, Santiago; MIEG, Adrés. Termoelectricidad, the energy balance. Técnica Industrial, v. 262, p. 62, 2006.

VERAS, J. et al. "An Automatic Thermal Cycling based Test Platform for Thermoelectric Generator Testing". In: Instrumentation and Measurement Technology Conference (I2MTC) Proceedings, 2015 IEEE International. IEEE, 2015. p. 1949-1953.

Block diagram in Labview ®

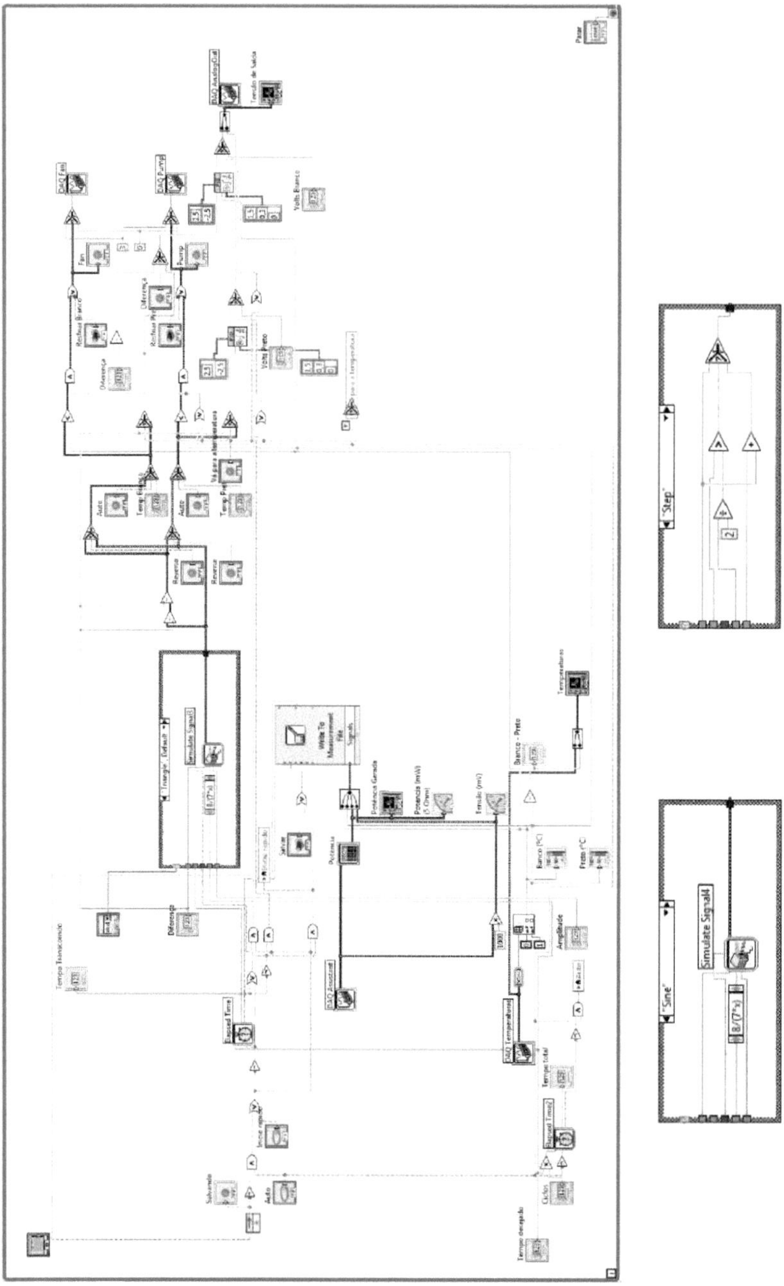

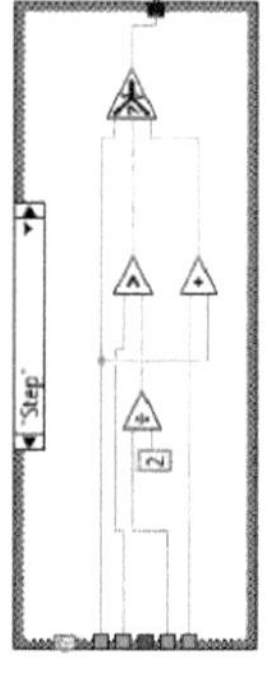
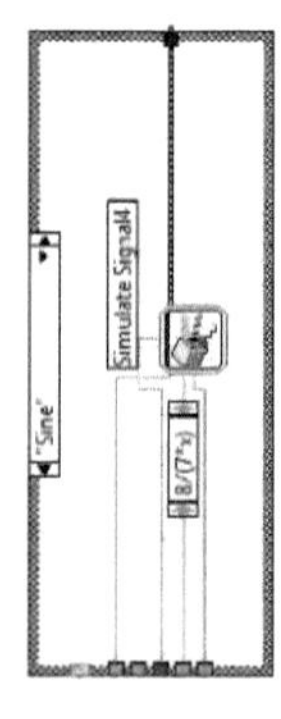

Printed by Books on Demand GmbH, Norderstedt / Germany